Playing Machines

By Betty Natelson

Printed in the United States of America

ISBN-13: 978-0-15-362464-3

ISBN-10: 0-15-362464-7

1 2 3 4 5 6 7 8 9 10 175 10 09 08 07

Introduction

Violins have been around for hundreds of years. They can sound sad, sweet, or lively. They are first-rate instruments, whether playing with a big orchestra or fiddling at a country dance.

A violin has to be made in just the right way to have a good sound. However, it is not made of anything fancy. An expert violin maker needs only pieces of the right wood, simple machines, and steady hands.

This is the story of a fictional violin maker, Ms. Morgan, and her music student, Erik. Ms. Morgan is starting to make Erik his first violin.

At her shop, Ms. Morgan has wood that she bought ten years ago. By now the sap and water have dried from the wood. This is important because vibrating wood gives the violin its sound. Dry wood vibrates easily, without cracking.

Ms. Morgan also has many tools. Some of the tools are simple machines. Others combine several simple machines into one device.

Cutting the Basic Pieces

Erik gazes around the shop, and he spots a finished violin on a shelf. Its back and belly look identical in size and shape. Together the back, belly, and ribs (sides) make up the body of the violin.

Ms. Morgan picks up a wooden mold the same size and shape as a finished violin. She lays the mold on top of a flat piece of maple wood, which she will use for the violin's back.

Holding the mold firmly, Ms. Morgan marks its outline on the maple wood. Then she lays the mold on a piece of spruce wood and outlines the violin's belly.

Ms. Morgan picks up a saw and cuts out the back and belly to the right general shape and thickness. The saw's blade is long and thin, with pointed teeth all along one edge. Each tooth is a wedge—a kind of simple machine. As the point of each tooth is pushed down, it splits the wood.

Erik watches while Ms. Morgan uses a knife to smooth the rough edges. The knife blade is thin along its cutting edge and thicker on its dull side. That makes it one long, thin wedge. It is so sharp that it cuts through even the hard maple wood of the violin's back.

Rounding the Back and Belly

Erik looks closely at the finished violin on the shelf. The back and belly are rounded, not flat. He wonders how his teacher will give his violin that graceful shape.

He looks back at Ms. Morgan. She is bending over the workbench, hands in motion, safety goggles covering her eyes.

Ms. Morgan uses a gouge to shape the violin's back. She shaves off wood from the violin's side up toward the center line, making the center higher. The violin's back and belly become curved. A gouge is another kind of wedge. Its wedge shape helps it cut into the wood.

Carving the F Holes

It's time to carve the F holes on the violin's belly. These are shaped like old-fashioned *f* letters.

Erik knows that a violin's sound doesn't come straight from its strings. The strings make hardly any sound at all. Instead, they cause vibrations inside the instrument. The F holes let out the sound so that people can hear it.

Ms. Morgan doesn't just stick a knife into the wood and start cutting curves. That might crack the wood. Also, the instrument's sound depends on carving the F holes correctly. Besides, if the F holes don't match each other, the whole violin will look odd.

Ms. Morgan takes a stencil and traces the F shape with a pencil. Next, she drills a hole in the middle of each of the circles she had marked at the top and bottom of the F. She uses a small hand drill and is very careful not to crack the wood.

The hand drill's bit is the metal part that cuts a hole in the wood. A drill bit is a pointed screw, another simple machine. The edge of the spiraling screw that winds up the bit is sharp, so it is another type of wedge.

A wheel-and-axle turns the bit. The wheel is the handle of the drill, which Ms. Morgan cranks around and around. That wheel turns a second wheel. The second wheel turns the drill bit, which is the axle.

Now Ms. Morgan picks up a small saw. She carefully cuts out the main part of the F holes, almost to the pencil lines. The little wedged teeth of the saw cut out the wood cleanly. Finally, with a thin, sharp knife, she finishes carving the F holes.

Purfling

The next day after school, Erik comes back. He knows that it's time for the purfling. *Purfle* is an odd-sounding word meaning "to decorate the edges or borders of something." On a violin, purfling is a line of wood set all around the borders of the back and belly. Purfling is mostly for decoration, but it does help make the violin a little stronger.

Ms. Morgan takes a thin, strong knife and cuts a channel all around the back piece. She cuts another line around the edge of the belly. Her steady hands carve the line perfectly.

Next, she picks up a chisel, a tool with the cutting part at its tip. A chisel blade is a wedge, with a flat back and an inclined plane at the front. Ms. Morgan chisels out the lines, making them into channels.

Earlier in the day, Ms. Morgan had cut thin strips of wood to use for purfling. Now she brushes hot glue on the wood, using the brush as a type of lever, another simple machine. The fulcrum is her wrist.

Carefully, Ms. Morgan lays the purfling into the channel. The hardest part is making the ends of the purfling meet exactly, so that the joining does not show. When Ms. Morgan finishes that job, she sits back with a sigh and a smile. She's done for today.

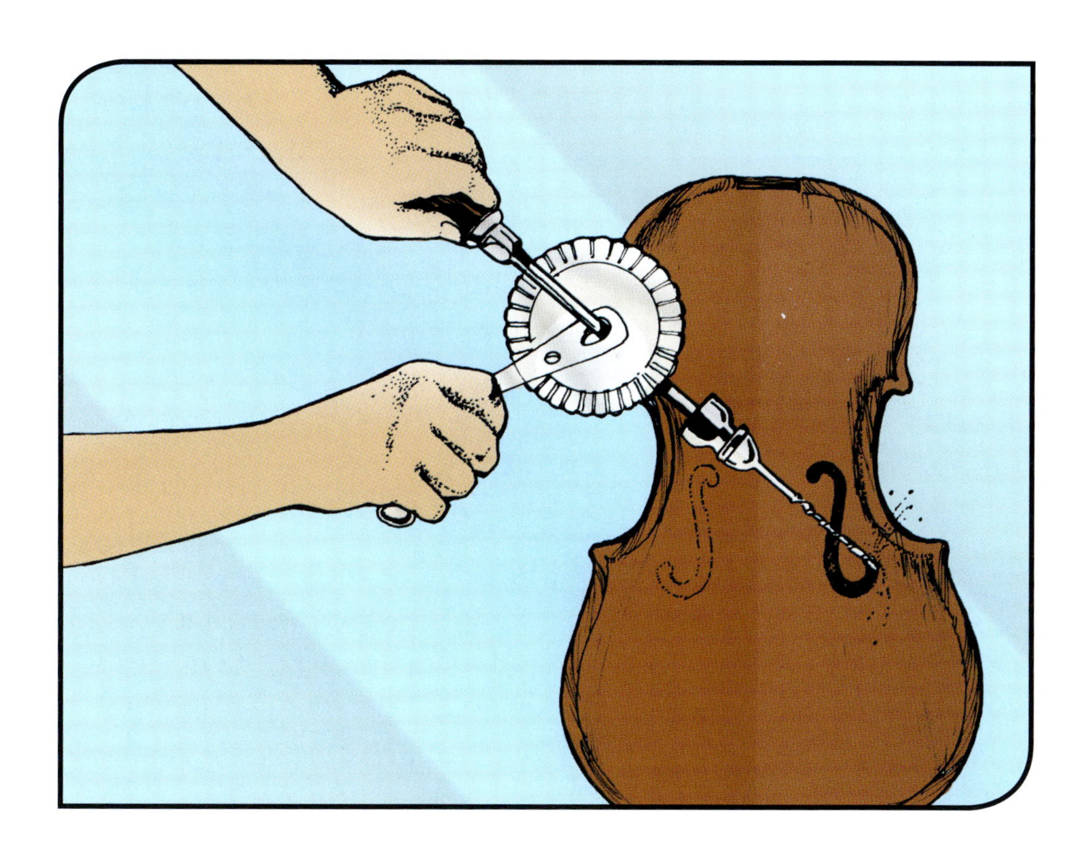

The Ribs

Erik watches while Ms. Morgan begins work on the ribs, the sides of the violin. She has cut some one-inch-wide strips of maple wood. Erik knows that she cannot just wind one rib all around the violin because wood doesn't bend that well. She needs a different strip for each curve.

Ms. Morgan uses a plane to make the maple strips even thinner. A plane's blade is another kind of wedge. On a plane, the blade is on the bottom. Ms. Morgan planes the ribs to make them thin. Then she soaks them in water until they are soft and flexible.

To shape the ribs, Ms. Morgan bends them around a hot metal bending iron.

Then she prepares the mold. She cuts wooden blocks and glues them inside the mold's C-shaped curves and inside the ends of the body. These blocks will be part of the finished violin, holding the ribs in place and protecting the violin from being crushed or falling apart.

Now Ms. Morgan wraps the shaped strips around the mold and its blocks. She glues the strips to one another, and they become the violin's ribs. She also glues them to the blocks.

Erik is worried that even with the blocks, the ribs may not be strong enough. He's right—they must be made stronger.

Ms. Morgan takes some willow wood and cuts out linings for the ribs. She shapes them on the hot bending iron and saws them to the right lengths. Finally, she brushes glue on them and sticks them to the inside of the ribs. Now the ribs will be strong enough.

To hold the linings to the ribs while the glue dries, Ms. Morgan clips on clothespins. She uses the kind with a spring in the middle. Each clothespin is made of two levers. The spring in the center acts as the fulcrum for both levers.

The Bass Bar

Erik runs into the workshop after school, a snack in his hand. He wants to see Ms. Morgan make the inside pieces for his violin.

First, Ms. Morgan carves and glues in the bass bar (BAYS•bahr), a strip of wood that helps the lower notes sound better and helps support the belly. That's important because when Erik plays his violin, the force of his bow on the strings will press down hard right on the belly's center.

The bass bar extends most of the length of the belly. Ms. Morgan has already chiseled and scraped for hours, carving the bass bar to match exactly the inside surface to which she will glue it.

Now she neatly brushes on glue and sticks the bass bar to the belly. It takes several clamps to hold the wood in place until the glue dries. A clamp combines several simple machines. To tighten the clamp, Ms. Morgan pushes a lever. The lever moves in a circle because it is the wheel of a wheel and axle. The axle is a fat screw. When Ms. Morgan turns the lever, she turns the screw. With each turn, the screw moves closer to a metal plate. The bass bar and belly are held together tightly between the screw and the metal plate.

Putting the Body Together

Erik comes to the shop feeling excited. Today his teacher will start putting together the violin's body. He watches Ms. Morgan set the ribs on the workbench. They are still around the mold so that they will keep their shape. Erik brings Ms. Morgan the back of the violin.

Ms. Morgan brushes glue onto the edges of the back and ribs, and then carefully lines them up. When she has the back and ribs together, she puts many small clamps all around the outside.

After the glue dries, Ms. Morgan is ready to take out the mold. This is a tricky job. One mistake and the ribs could crack. Gently she taps the blocks with a hammer to loosen them from the mold. Then she grasps the mold by its holes and lifts it out. Ms. Morgan uses the hammer as a type of lever. Because she is tapping gently, her wrist is the fulcrum. If she were using her whole arm to hit harder, her elbow would be the fulcrum.

Erik knows that it's time to put the belly on his violin. Ms. Morgan brushes glue onto the edges of the belly and ribs and expertly joins them. Then she clamps them together and waits for the glue to dry.

Making the Neck

Erik is eager to see his teacher carve the neck for his violin. With the neck and the strings, he'll be able to play it.

Using a stencil and scriber, Ms. Morgan marks the neck's outline on a piece of maple wood. A scriber is a tool that cuts a thin line with its sharp metal point. It is another type of wedge.

With the line cut by the scriber as a guide, Ms. Morgan saws a rough outline of the neck. Then she carves it with her knife and other tools until it looks just right.

The hardest part to make is the scroll at the end. Ms. Morgan spends a very long time cutting that part. The scroll is just decoration, but it is part of the violin tradition.

Next to the scroll, Ms. Morgan makes the pegbox. Using her hand drill, she puts four holes in the pegbox. Then she carves pegs to go into the holes. Later she will wind strings around the pegs.

At last, Ms. Morgan glues the neck to the body. She uses a clamp to hold the two pieces together until the glue dries. She and Erik know that it won't be long until Eric will be able to play his new violin.

Ms. Morgan smooths the scroll with a knife.

Varnishing

Erik watches while Ms. Morgan brushes varnish on every bit of the wood. Varnish gives the wood a hard, shiny coat, both to protect it and to give its vibrations a better sound. Erik can clearly see the grain of the wood through the varnish. Ms. Morgan's varnish brush is a lever. Her wrist is the fulcrum.

A week later, Ms. Morgan checks the violin. The varnish has dried. She rubs it smooth before putting on the next coat. Some violin makers put fifteen coats of varnish on the violins they make.

The Final Pieces

Erik hurries to the violin shop right after school. Today Ms. Morgan will begin the finishing touches on his violin.

Ms. Morgan already has black ebony wood for the fingerboard. That's where Erik's left-hand fingers will go when he presses the strings.

Using a steel scraper, Ms. Morgan carves the strip of ebony to be exactly right for this violin. Then she brushes on glue and places the fingerboard on the neck of the violin. She clamps the two parts together to dry.

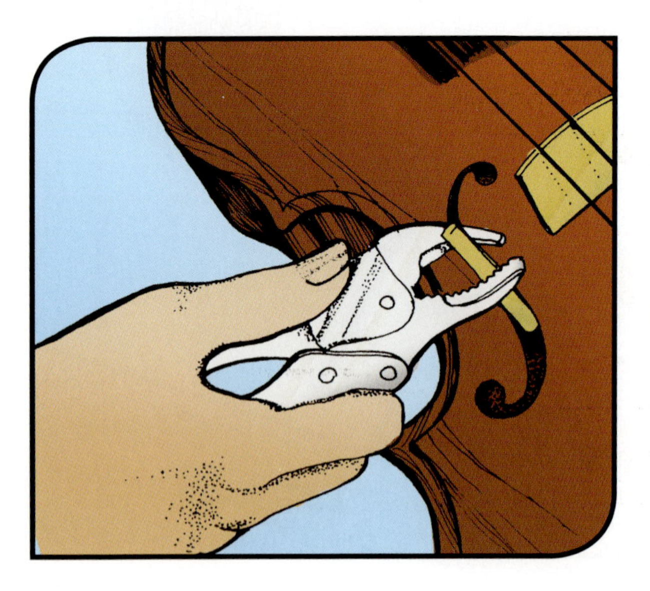

The pliers used to set the soundpost are made up of two levers. The screw in the middle is the fulcrum for both levers.

Later, Ms. Morgan sets up a bridge between the two F holes. In addition to holding up the strings, the bridge carries vibrations to the wooden body of the violin. This makes possible the instrument's special sound.

Next, Ms. Morgan slips the soundpost through an F hole and braces it between the back and the belly. Erik knows that this little pole will carry vibrations to the back, belly, and throughout the violin. It will also help hold up the belly when the strings vibrate.

Ms. Morgan uses a tool that looks like a thin pair of pliers. Like the clothespins she used earlier, this tool combines two levers, with the fulcrum in the center.

Finally, Ms. Morgan adds the chinrest and the tailpiece.

The Strings

Erik fidgets impatiently while Ms. Morgan stretches each of the four strings from the tailpiece to a peg near the end of the neck. Each string has a different thickness. The thicker it is, the lower the sound it makes.

Ms. Morgan winds each string around its peg. When he's ready to play, Erik will turn the pegs to tighten or loosen the strings to make them sound at the right pitch. Each peg is a wheel and axle. The wheel is the knob, and the axle is the post of the peg.

Playing the Violin

At last, Erik picks up his violin. With his left hand, he holds its neck. His fingers press down the strings.

Erik moves the bow with his right hand. He does not push the bow down hard against the strings. It is not a saw. Instead, he rests it lightly on the strings so that it glides over them. He has put rosin, a sticky powder made from tree sap, on the hairs of his bow. This makes the bow hairs drag just enough to make the strings vibrate. Those vibrations move through the violin's wooden body and come out the F holes as music.

Erik notices one last simple machine. When he plays the violin, his arm is a lever holding up the bow, and his shoulder is a fulcrum.

The violin has an excellent sound. Ms. Morgan is pleased to hear its music, and she is proud to hear her student play so well.